1 MONTH OF
FREE
READING

at
www.ForgottenBooks.com

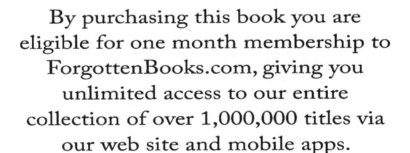

By purchasing this book you are eligible for one month membership to ForgottenBooks.com, giving you unlimited access to our entire collection of over 1,000,000 titles via our web site and mobile apps.

To claim your free month visit:
www.forgottenbooks.com/free911730

ISBN 978-0-266-93189-8
PIBN 10911730

This book is a reproduction of an important historical work. Forgotten Books uses
state-of-the-art technology to digitally reconstruct the work, preserving the original format
whilst repairing imperfections present in the aged copy. In rare cases, an imperfection in
the original, such as a blemish or missing page, may be replicated in our edition. We do,
however, repair the vast majority of imperfections successfully; any imperfections that
remain are intentionally left to preserve the state of such historical works.

BOSTON UNIVERSITY

GRADUATE SCHOOL

Thesis

EVOLUTION OF THE CAROTID SINUS

AND ITS HOMOLOGUES

by

George P. Fulton

(S. B., Boston University, 1936)

submitted in partial fulfilment of the

requirements for the degree of

Master of Arts

1938

AM
1938
+o

Introduction

The carotid sinus and its homologues are part of a physiological mechanism which has three important functions. First,
this mechanism reflexly regulates the circulation. Second, it
reflexly regulates the respiration. Third, it reflexly main-
tains the normal tone of the heart, the vasomotor system and
the respiratory system.

The chief homologue of the carotid sinus is the arch of the
aorta, which is innervated by the aortic depressor nerve. Cyon
and Ludwig (1866) first described this nerve as a slender fila-
ment accompanying the vagus and the cervical sympathetic nerve
in the rabbit. When the central end of the aortic depressor
nerve was electrically stimulated, there occurred a decrease in
the heart rate and a fall in the general arterial blood pressure. When the peripheral end of the same nerve was stimulated
no such effects occurred. These observations have been confirmed many times and are now repeated as a routine laboratory
procedure in physiology.

Further experimental work has established the fact that the
aortic arch contains receptors which are stimulated by changes
in the blood pressure and by the chemical content of the blood.
The receptors responding to pressure are called pressoreceptors,
and those responding to chemical conditions are called chemoreceptors. In general, stimulation of either set of receptors
results in three types of reflex. First, afferent impulses

pass to the cardiac center and are relayed as efferent impulses
to the heart. Second, afferent impulses pass to the vasomotor
center and are relayed as efferent impulses to the blood vessels. Third, afferent impulses pass to the respiratory center
and are relayed as efferent impulses to the muscles involved in
respiration. The afferent pathways consist of the sensory neurones which compose the aortic depressor nerve. The efferent
pathways consist of the motor neurones, autonomic and cerebrospinal, which innervate the heart, blood vessels and respiratory muscles. (Figure 1).

The carotid sinus was described in man as early as 1811 by
Burns and has been consistently observed in the vertebrates.
It is a slight dilation of the internal carotid artery, near the
bifurcation of the common carotid. Some of the early anatomists believed that the carotid sinus was the result of a pathological condition. Others thought that the internal carotid
had undergone retrogression and that the dilation was merely a
remnant. Schafer (1877) suggested the true origin of the carotid sinus when he speculated that it might be a remnant of the
third branchial arch. But the significance of this embryological origin has only recently been pointed out.

In 1866, the same year in which the aortic depressor mechanism was discovered, Tschermak called attention to a puzzling
phenomenon which had been noticed previously by a few observers.
Pressure with the finger applied to a man's neck at the region
of the bifurcation of the common carotid artery resulted in a

decrease in heart rate. Tschermak believed the slowing to be
due to direct mechanical stimulation of the vagus, and named
the phenomenon the "Vagusdruckversuch". This idea persisted
for many years and the experiment was widely used clinically as
a test for vagal excitability.

A few years later François-Franck (1870) performed a series
of experiments leading to the concept that control of the blood
supply to the head is governed by the direct effect of alterations in pressure on the centers in the medulla. These experiments were confirmed by many authorities (Hedon, 1910; Heymans
and Ladon, 1925; Anrep and Segall, 1926). The concept of direct
effect on the centers became so firmly established that investigators lost sight of the possibility of a reflex with its receptors within the vessels of the head itself. This is surpri-
zing considering the previous discovery of the aortic depressor
reflex and the recognition of its importance in reflexly regulating the blood supply to the distribution of the aorta.

Furthermore, the carefully performed experiments of Siciliano (1900) and Pagano (1900) which presented sound evidence for
a reflex mechanism went unnoticed. Pagano concluded, as a result of clamping and releasing the common carotid artery and its
efferent branches, that the region nearest the bifurcation was
the source of a reflex which altered the heart rate upon change
of mechanical pressure within this region. Siciliano recognized
the importance of this reflex in supplying an adequate amount
of blood to the head and brain. However, their experiments

were contested and ignored.

In 1924, H. E. Hering demonstrated that stimulating either
the carotid sinus itself or the carotid sinus nerve resulted in
a reflex slowing of the heart and a fall in arterial blood pressure. "Ces faits expérimentaux démontrèrent donc d'une manière
péremptoire la sensibilité réflexogène du sinus et du nerf inter-
carotidien et mirent en evidence d'autre part que le *Vagusdruck*
de Tschemak n'était que le 'Carotissinusdruck' de H. E. Hering--
E. Koch". (Heymans, Bouckaert, and Regniers, 1933)

Further experimental work has established the fact that the
carotid sinus contains both pressoreceptors and chemoreceptors.
When these receptors are stimulated, reflexes regulating the
heart, the blood vessels and the respiratory muscles occur just
as in the case of the receptors in the aortic arch. (Figure 1).

Closely associated with the carotid sinus is a small globose
structure, which was described by Haller in 1743, and which has
since undergone numerous changes in nomenclature corresponding
with the numerous speculations as to the origin and function.
It is now referred to as the carotid body or glomus caroticum
and is known to be a part of the carotid sinus mechanism. De
Castro (1926), (1928) made the anatomical studies which served
as the basis for the physiological investigations which have
shown that the chemoreceptors are located in the carotid body.

The significance of the carotid sinus and its homologues is
best understood when they are examined from an evolutionary
point of view, Koch (1931) stated that the parts of the circu-

latory system in which the pressoreceptors are located, are the adult survivals of the primitive arterial system of the embryo. Lutz and Wyman (1932) found a cardio-inhibitory reflex in the elasmobranch. The receptive area in this instance was quite widespread among the gill vessels. Consequently the investiga- tors believed that during the evolutionary process the receptive areas became delimited to the carotid sinus and aortic arch, both of which have been derived from the primitive gill arch vessels.

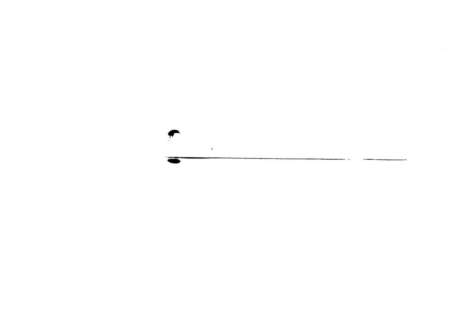

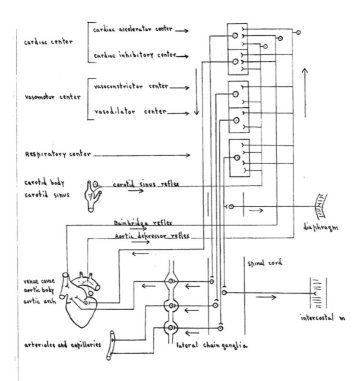

cardiac center
— cardiac accelerator center →
— cardiac inhibitory center →

vasomotor center
— vasoconstrictor center →
— vasodilator center →

Respiratory center _____→

carotid body
carotid sinus
carotid sinus reflex
→

Bainbridge reflex
Aortic depressor reflex

venae cavae
aortic body
aortic arch

arterioles and capillaries

lateral chain ganglia

diaphragm

spinal cord

intercostal m

Functional Diagram of Carotid Sinus and Homologous Reflex
Mechanisms for Cardio-vascular and Respiratory Regulation

Fig 1

Part I.   Reflex Responses to Changes in Blood Pressure.

The pressoreceptors, which are activated by changes in the
blood pressure, are located both in the arch of the aorta and
in the carotid sinus. Those pressoreceptors found in the arch
of the aorta are the terminations of the aortic depressor nerve.
Those found in the carotid sinus are the terminations of the
carotid sinus nerve.

The aortic depressor nerve is a branch of the vagus. In the
rabbit, it is a slender nerve which accompanies the vagus and
cervical sympathetic. In the dog, it does not occur as a distinct nerve bundle in the sheath which contains the vagus and
cervical sympathetic, but in the vicinity of the heart it branches from the vagus. In the cat, about every fourth aortic depressor can be separated from the trunk of the vagus. The aortic depressor nerve is paired. The left branch innervates the
aortic arch and the right branch innervates the right subclavian artery.

The carotid sinus nerve is a branch of the glossopharyngeal.
In the dog, both the vagus and cervical sympathetic send nerve
fibers to the bifurcation of the common carotid artery, but the
glossopharyngeal is the main pathway for the carotid sinus reflexes. Code, Dingle and Moorehouse (1936) obtained evidence
that afferent fibers from the carotid sinus in the dog, reached
the reflex centers by way of the cervical sympathetic nerve and
spinal cord. They considered that this pathway was relatively

unimportant.

The carotid sinus and aortic arch are both adult survivals of the primitive branchial arches. They are both innervated by nerves which are embryologically and comparatively the nerves of the branchial arches.

Tello (1924) worked out the development of the nerve endings in the aorta of the mouse and rabbit. He found that the aorta was at first in contact with the superior laryngeal nerve and the nodose ganglion of the vagus. In the 10 millimeter stage, the aortic arch became separated from the superior laryngeal nerve, but remained connected with the vagus by fibers which formed the principal branch of the aortic depressor nerve. Tello found no fibers within the heart.

Nonidez (1935a) investigated the development of the aortic arch pressoreceptors in the chick. He found the aortic depressor nerve on the right side but not on the left. He stated that this condition existed because in birds the right fourth branchial arch artery persists to become the aortic arch, whereas the left fourth branchial arch artery disappears and does not persist as subclavian artery. Nonidez found an accessory aortic depressor nerve arising directly from the ganglion nodosum.

That the carotid sinus of mammals is a derivative of the third branchial arch has long been known to anatomists. In the bird, Muratori (1937) stained pressoreceptor endings in the artery which is usually referred to as the common carotid.

lar system was homologous with the carotid sinus of mammals and
that the term, common carotid, was a misnomer. He stated that
this portion of the arterial system was embryologically the internal carotid artery. Ask-Upmark (1935) found that two seg-
ments of the carotid artery in the bird are innervated. The
upper segment was medial to the mandibular angle and the lower
segment was at the level of the thyroid artery.

The pressoreceptors are quite characteristic in appearance.
De Castro (1926, 1928) stained these endings in the carotid sinus of certain species of mammals, including man, by using a
silver stain and also methylene blue. He found numerous terminal arborizations ending in reticulated swellings. These were
located in the adventitia and were of two types, diffuse and
limited. They were situated between the elastic and collage-
nous fibers. De Castro believed that their position accounted
for their response to changes in the hydrostatic pressure. He
found these endings only in the region of the carotid sinus.

Nonidez (1935) stained the pressoreceptor type of nerve endings in the adventitia of the aortic arch in the young cat,
rabbit and guinea pig. He used a modification of the silver
nitrate method developed by Ramón y Cajal. He stated that the
outstanding feature of these receptors was their swellings or
enlargements in which neurofibrils were easily detected. (Figure 7, page 38).

Three types of reflexes occur when the pressoreceptors in

the aortic arch and carotid sinus are activated by an increase
in blood pressure. These are cardiac reflexes, vasomotor reflexes and respiratory reflexes. C. Heymans and co-workers are
largely responsible for the fundamental findings concerning
these reflexes. They have published the results of their findings together with an extensive review of the literature in the
monograph, "Le Sinus Carotidien et la Zone Homologue Cardio-
aortique", G. Doin, Paris, (1933).

A.   Cardiac Reflexes.

The exact pathway of the impulses originating in the aortic
pressoreceptors and resulting in cardiac slowing has been worked
out by physiological methods applied chiefly to dogs. Electrical stimulation of the aortic depressor nerve after cutting the
vagus innervation of the heart produces only a slight slowing
of the heart. This bradycardia is due to an inhibition of accelerator tone, for the reflex is reciprocal. Opinions differ
as to whether the cardiac reflex is homolateral or heterolateral. The strength of stimulus necessary to bring about the heterolateral response in those species reported to possess a heterolateral reflex is much greater than the strength required
to bring about a homolateral response. In addition to the bradycardia there is a decrease in the output of the heart, and a
fall in blood pressure results.

The adequate stimulus for the heart reflex is a rise in
blood pressure within the aorta itself. This has been demonstrated conclusively by Heymans and collaborators by means of

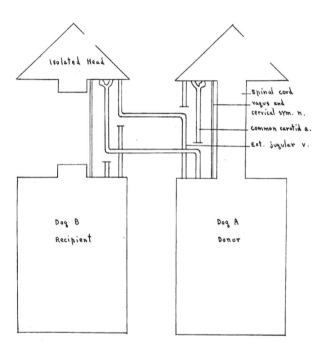

Isolated Head

spinal cord
vagus and cervical sym. n.
common carotid a.
Ext. jugular v.

Dog B
Recipient

Dog A
Donor

Isolated   Head   Technique

Fig. 2

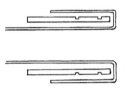

Payr Cannula

Fig. 3

the isolated head technique. Two dogs, A (the donor) and B
(the recipient) were anaesthetized and placed side by side.
The carotid artery and the external jugular vein on each side
in both animals were exposed then sectioned between two ligatures. Then the cardiac end of each carotid artery of dog A
was connected with the cephalic end of each carotid artery of
dog B. The cephalic end of each jugular vein of dog B was connected with the cardiac end of each jugular vein of dog A. (Figure 2). The connections were established by using arteries taken from a third dog. A direct intima to intima anastomosis,
which rendered the use of an anticoagulant unnecessary, was effected by using a Payr Cannula. (Figure 3). A tracheal cannula was inserted into the trachea and the blood pressure was recorded by a mercury manometer connected with a cannula inserted
into the femoral artery. Then the entire head of dog B was severed from its body except for the vagus nerves and cervical sym-
patnetic nerves. By this ingenious method, the head of dog B
was supplied with blood from dog A.

By means of the isolated head technique it was found that an
increase in blood pressure in dog B, brought about by an intravenous injection of defibrinated blood or adrenalin, produced a
slowing of its heart. On the other hand, a decrease in blood
pressure, brought about by bleeding, produced a speeding up of
the heart. These responses no longer appeared when the vagus
nerves were cut.

The exact pathway of the impulses originating in the carotid
sinus pressoreceptors has been worked out also by electrical

stimulation experiments. This heart reflex is also reciprocal
in nature. Bradycardia is produced both by activation of the
vagal inhibitory center and inhibition of the accelerator center. There is no general agreement as to whether the reflex is
homolateral or heterolateral. In the dog, Heymans found that
the heart reflex was homolateral and attributed the contrary
findings of others to be due to an excessively intense stimulation which reflexly inhibited the accelerator tone and slowed
the heart without having involved vagal inhibition. However
there may be a species difference, since Hoff and Nahum (1935)
found that the carotid sinus reflex was bilateral in monkeys.
Also, Lutz and Wyman (1932) found that the heart reflexes
brought about by increasing the pressure of heparinized blood
which was used to perfuse the gill vessels were both unilateral
and bilateral.

Bronk, Ferguson and Solandt (1934) investigated the reflex
pathway by means of a vacuum tube amplifier and oscillograph.
They recorded the action currents which accompanied the passage
of nerve impulses in one of the small nerve fibers from the
stellate ganglion to the heart in the cat. When the perfusion
pressure in the isolated carotid sinus was raised, discharges
decreased until, at 125 to 150 millimeters of mercury, there
was a complete inhibition of sympathetic impulses. The duration
of the inhibition was a function of the endosinusal pressure.
The period of inhibition lasted for some seconds, after which
an escape occurred perhaps due either to an adaptation of the

nerve endings in the carotid sinus or to an adaptation in the nerve centers. The investigators obtained evidence that the reflex inhibition of accelerator tone was bilateral. Section of both aortic depressor nerves prolonged the inhibition.

Various methods for producing a stimulation more physiological than an electric current, have been devised. The first of these methods was developed by E. Moissejeff (1926) and employed in modified form by Heymans. It may be called the blind pouch technique. All the efferent arteries from the carotid sinus in a dog are tied off, care being taken not to interfere with the innervation. Then the common carotid is sectioned between two ligatures and the cephalic end is connected with a burette containing the perfusion fluid. A mercury manometer connected to a cannula which is inserted in the femoral artery or carotid artery on the opposite side records the heart rate and blood pressure. The height of the burette can be varied in such a way as to produce any desired degree or pressure within the isolated carotid sinus. (Figure 4).

Another method designed to produce a stimulus even more physiological was developed by Heymans. This method is somewhat similar to that of the isolated head. It may be called the isolated carotid sinus technique. Two dogs, A (the donor) and B (the recipient) are anaesthetized and placed side by side. The carotid sinus of dog B is exposed and the efferent arteries leading from it are tied off. The external jugular vein and common carotid artery of dog A are exposed and cut between two

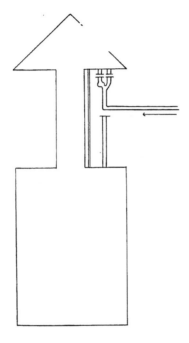

Blind Pouch Technique

Fig. 4

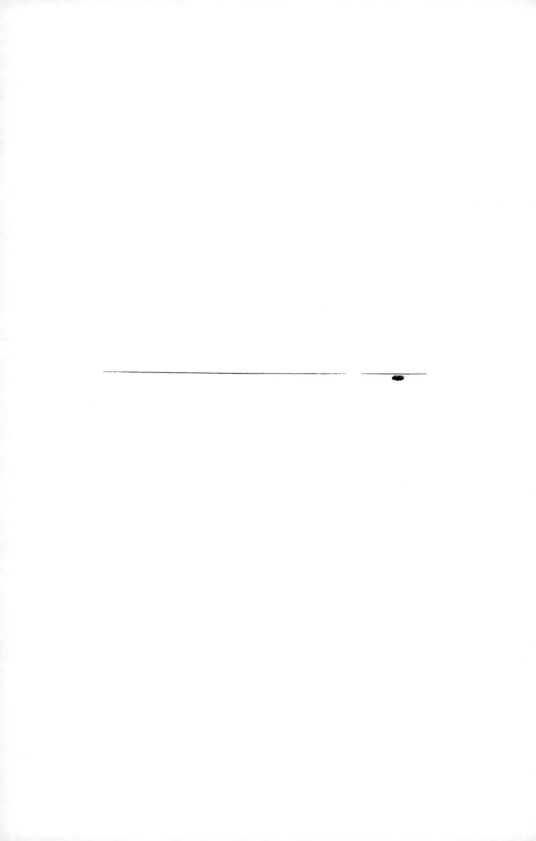

ligatures. Then the cardiac end of the carotid of dog A is connected with the cephalic end of the carotid of dog B. The cardiac end of the external carotid artery of dog B is connected
with the cardiac end of the jugular vein of dog A. (Figure 5).
As in the case of isolated head technique, the connections are
made by using arteries taken from a third dog. Thus the carotid
sinus is completely isolated except for its nerve supply.

By using such experimental techniques as these, investigators
have established beyond doubt that a rise in blood pressure in
the carotid sinus produces a reflex slowing of the heart. However, some investigators (Nash, 1926; Anrep and Segall, 1926)
still believed that blood pressure exerted a direct effect on
the cardiac center as well as a reflex effect on the carotid sinus. The following four experiments performed by Heymans and
co-workers have refuted such beliefs. First, the occlusion of
both common carotid arteries resulted in tachycardia. But,
when the carotid sinuses were denervated, no modifications of
heart rate occurred. Second, the occlusion of both internal
carotid arteries resulted in bradycardia. If the effect had
been on the medullary centers, a rise in blood pressure and increased frequency of the heart would have been expected. Third,
the occlusion and release of the vertebral arteries did not produce any change in heart rate. Fourth, the prolonged hypertension produced by injection of Ringer's solution or defibrinated
blood had no effect upon the nerve centers after denervation of
the carotid sinuses.

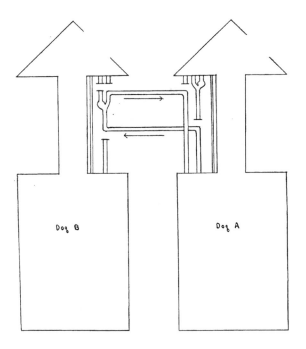

Isolated Carotid Sinus Technique

Fig. 5

Additional evidence was obtained by using the isolated head
technique which has been described above. It was found that hypotension in the isolated but perfused head of dog B, which
could be brought about by either bleeding dog A, or stimulating
its carotid sinus or aortic depressor nerves, resulted in tachycardia in the trunk of dog B. Hypertension in the isolated head
of dog B, which could be effected by injection of adrenalin into
dog A, resulted in bradycardia in the trunk of dog B. When one
carotid sinus was denervated, only the innervated side responded.
When both carotid sinuses were denervated, no responses occurred.

In another experiment Heymans perfused the isolated head of
dog B with blood from dog A and the isolated carotid sinus of
dog B with blood from dog C. The nerve supply to the isolated
carotid sinus was left intact, but the opposite carotid sinus
was denervated. It was then shown that modifications of pressure in the same animal produced reflex effects at the level of
the carotid sinus but did not influence the vagus center.

In these experiments it was not possible to vary the pressure in the isolated carotid sinus at will. Hence, Heymans and
his collaborators sometimes found it convenient to perfuse the
isolated carotid sinus with a pump which created a pulsatile
flow. This was of particular value in studying the effects of
step-by-step increases or decreases in pressure. In the following chart which is reproduced from the monograph by Heymans,
the variations, in percent, of the arterial blood pressure are
plotted against the corresponding perfusion pressure used to

elicit the blood pressure variations. An analysis of this grap
shows the following three facts.

First, below a pressure of 50 millimeters of mercury in the
carotid sinus, every increase of endosinusal pressure resulted
in an increase of arterial pressure. And every decrease of end
sinusal pressure resulted in a decrease of arterial blood pressure. Second, above a pressure of 50 millimeters of mercury in
the carotid sinus, every increase of endosinusal pressure resul
ed in a decrease of arterial pressure. And every decrease of
endosinusal pressure resulted in a rise of arterial blood pre-
sure. Third, a maximum sensitivity occurred between 85 to 110
millimeters of mercury.

Heymans interpreted the above three facts in the following
way. A rise in blood pressure when the intracarotid pressure is
near the normal level, 85 to 110 millimeters of mercury, slight
distends the arterial wall and activates the receptor endings,
resulting in hypotension and bradycardia. At very high intracarotid pressures, however, the receptors are maximally excited
and no further increase in pressure ca. augment the reflex responses. On the other hand, at very low intracarotid pressures
each decrease in pressure serves to collapse the arterial wall
of the carotid sinus and further activate the receptors, result
ing in hypotension and bradycardia. Each rise in pressure serv
to lessen the tension on the receptors, resulting in a decrease
of the reflex responses. Modifications of pressure as slight
as 1 to 2 millimeters of mercury above or below the normal bloo

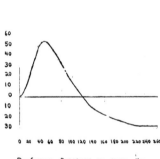

Perfusion Pressure in mm Hg.

pressure evoke the reflex responses.

Bronk and Stella (1932, 1932a, 1935) studied the nature of the nervous discharge from single pressure receptors in the carotid sinus. They cut the carotid sinus nerve in the rabbit at its origin from the glossopharyngeal. Then they cut all the fibers of the carotid sinus nerve near its distribution except one or a few. They placed electrodes on the main stump of the nerve and recorded the action potentials by means of a vacuum tube amplifier and Matthews oscillograph. Impulses occurred at a regular frequency in the manner which is characteristic of most sense organs. The receptors did not become adapted readily but corresponded, in respect to adaptation, with muscle spindles rather than touch receptors, which have a brief duration. It was also found that a volley of afferent impulses occurred with each arterial pulse. Thus, each beat of the heart itself seems to initiate impulses which help maintain the normal cardiac and vasomotor tone.

Bronk and Stella also perfused an isolated carotid sinus and recorded the action potentials from one or several end organs during varying pressures from 40 millimeters of mercury to 200 millimeters of mercury. Over this range there was an increased discharge of two and one-half times. Not all the endings had the same threshold. Thus an increase in pressure increased the number of afferent impulses from the carotid sinus both by increasing the frequency of discharge from a single end organ and by increasing the number of end organs acting. Furthermore,

they found that a rapid fall in pressure such as from 120 millimeters of mercury did not elicit a frequency of discharge characteristic of constant perfusion at that low pressure. But the
discharges were completely inhibited for as long as 18 seconds,
then they began again at a new frequency. The authors believed
that this phenomenon accounts for the decreased discharge from
the carotid sinus which accompanies a fall in diastolic pressure.
Nonidez (1935) believed that the great variety of sizes and
shapes of the pressoreceptor endings accounts for the difference
in threshold, the more delicate terminations having the highest
threshold.

Bronk and Stella (1935) tested the effects of oxygen lack and
carbon dioxide excess on these receptors in the isolated carotid
sinus by perfusing it with blood of varying chemical content.
There was no increase in the frequency of the action potentials.
Thus it was concluded that the receptors in the carotid sinus
are highly insensitive to changes in the chemical content of the
blood. The authors pointed out that this is not evidence against
the role of the carotid sinus in the regulation of respiration
but suggests that there is a different set of endings influenced
by changes in the chemical content of the blood.

Winder (1937) investigated the carotid sinus reflexes in an
animal with intact aortic depressor nerves. Upon eliciting the
carotid sinus reflex, he found that bradycardia was poorly maintained, due to the compensatory action of the aortic depressor
reflex. The experiments performed by Winder indicate that in

the intact animal there is a tendency for the aortic depressor nerve to reverse the action of the carotid sinus nerve. Thus, bradycardia due to increased pressure in the carotid sinus may become tachycardia due to the resulting decreased pressure in the aortic arch.

Another of the reflexes regulating the heart is the Bainbridge reflex. Bainbridge (1915) found that increasing the pressure in the right auricle of the heart by injecting fluid or blood through the jugular vein brought about an increase in the heart rate. This acceleration disappeared when the vagus nerve was cut, hence it was due to a reflex with afferent pathway in the vagus nerve. The speeding up of the heart was found to be due to a diminution of vagal tone and an increase of accelerator tone. There was no evidence of increased output of the adrenal gland. There was no evidence that this reflex was concerned in either vasomotor or respiratory responses. (Figure 1).

McDowall (1935) found a branch of the vagus nerve given off to the heart in the cat at the level of the right subclavian artery. He found that stimulation of the cranial end of this branch brought about an acceleration of the heart and an increase in blood pressure when the main trunk of the vagus was intact. Since the effects were no longer produced when the vagus trunk was severed, they may be regarded as reflex in origin. It is possible that this is the nerve concerned in the mediation of the Bainbridge reflex.

The existence of this reflex has been debated by physiolo-

gists for some time. However, Nonidez (1937) has provided the
anatomical confirmation by staining nerve endings of the pressoreceptor type in the heart of the newborn rabbit, cat and dog.
He found these terminations in the walls of the superior vena
cava, inferior vena cava and coronary sinus at their junctions
with the right atrium and in the proximal portions of the pulmonary veins. The endings occurred in the subendothelium and
among the myocardial bundles which extend into the walls of the
great veins as far as the pericardium.

The importance of this reflex in protecting the heart is ap-
parent. Thus, during strenuous muscular exercise, the venous
inflow to the right side of the heart is greatly increased by
the pushing of blood past the valves in the veins due to the mechanical effects of muscular contractions. In such a situation
the Bainbridge reflex prevents dilation of the heart. It also
assures the heart and other tissues of an adequate blood supply
by working in opposition to the carotid sinus and aortic depressor mechanisms. Thus, if the reflexes from the arterial
side, particularly the aortic depressor, have slowed the heart,
the resulting increased venous filling exerts a pressure adequate to elicit the Bainbridge reflex.

B.   Vasomotor Reflexes

Cyon and Ludwig in their original experiments noticed that
a fall in blood pressure occurred, as well as a slowing of the
heart. Numerous experiments have established the fact that the
vasoconstrictor and vasodilator centers are influenced recipro-

cally by impulses arising from both the arch of the aorta and
the carotid sinus. There is some evidence that veins (particularly those in the hind leg, intestine, and mesentery) and
capillaries are reflexly involved in the vasomotor responses, as
well as arterioles.

The efferent pathway of the vasomotor portion of the carotid
sinus reflex has recently been investigated in the cat by Thomas
and Brooks (1937). They removed the thoracic and abdominal sympathetic chain by severing the connections between the spinal
cord and the postganglionic sympathetic fibers. The vagus and
sympathetic trunks were cut in the lower cervical region to
abolish the heart reflexes and influences of the aortic depressor mechanism. They found that an abrupt occlusion and release
of the common carotid arteries in sympathectomized animals re-
suited in an average rise in pressure of 11.1 millimeters of
mercury in 11 cats as compared with an average rise in pressure
of 86.7 millimeters of mercury in 18 normal cats. When they repeated the occlusion and release experiments, after cutting the
carotid sinus nerve, only a slight rise occurred which was of
the order of 11 millimeters of mercury. This suggested that the
rise in the sympathectomized cats with innervated carotid sinuses
was not reflex but mechanical in nature. Experiments in which
the isolated carotid sinus was perfused with Ringer's solution
at different pressures further demonstrated that reflex alterations did not occur in sympathectomized animals. It was thus
concluded that the efferent pathway of the vasomotor carotid

sinus reflexes must be in the sympathetic chain.

The vascular area involved in the vasomotor reflexes elicited at the carotid sinus and the aortic arch is quite extensive.
The evidence for the vascular areas involved has resulted from
a large number of experiments by many investigators. The results of these experiments have been exhaustively summarized by
Heymans (1933) and McDowall (1935). From their review of the
literature, the following table has been constructed. The
asterisk indicates evidence of a carotid sinus or aortic depressor vasomotor reflex. The number beside the asterisk indicates roughly the number of investigators who have reported the
reflex.

The secretion of adrenin is augmented during vasoconstriction and diminished during vasodilation. Heymans and his collaborators (1933) demonstrated this by the following method.
The carotid sinus of dog B was isolated and perfused with blood
from dog A. Dog B was epinephrectomised on the left side and
its adrenolumbalis vein was anastomosed with the cardiac end of
the right external jugular vein of dog C. Dog C was epinephrectomised on both sides and the volume of its spleen was recorded
by means of a plethysmograph. Thus, the output of adrenin in
dog B was indicated by a shrinkage of the spleen of dog C. It
was found that a fall in blood pressure in the carotid sinus of
dog B, produced by compression of the perfusing artery, resulted
in an increased output of adrenin as evidenced by a marked constriction of the spleen of dog C. A rise in blood pressure in

23

Arterial Areas Involved in the Vasomotor Response to
Stimulation of the Receptors in the Carotid
Sinus and the Aortic Arch

Location of Arterioles	Carotid Sinus Reflex	Aortic Depressor Reflex
Cerebrum	* ?	
Head	* 1	* 2
Ear		* ?
Nose	* 1	* 2
Tongue		* 2
Submaxillary gland		* 2
Retina	* 1	
Extremities	* 4	* 3
Skin	* 3	* 3
Muscles	* 1	* 3
Heart	* 1	* 1
Lungs	* 1	* 1
Thyroid	* 1	* 1
Liver	* 1	
Spleen		* 1
Intestine	* 3	*10
Kidney	* 1	* 6
Penis	* 1	* 1

the carotid sinus of dog B, produced by injecting adrenalin into dog A, resulted in a decreased output of adrenin as evidenced by a dilation of the spleen of dog C.

Even before the discovery of the carotid sinus mechanism, the
dilatation at the origin of the internal carotid led to speculation concerning its possible relation to the blood supply of the
brain. As stated in the introduction, Siciliano recognized the
importance of reflexes elicited within the carotid arteries, in
protecting the brain against anemia. After Hering published the
results of his experiments, investigators began to stress the
protective relation of the carotid sinus to the cerebrum. Further evidence of this protective function was presented by de
Castro (1928) who found that the carotid sinus in the cow was
situated at the origin of the occipital artery which is the main
source of arterial blood to the brain in that animal.

Ask-Upmark (1935) investigated the importance of the carotid
sinus in relation to the cerebral blood supply. He did this by
a comparative study of the anatomy of the carotid sinus in as
many species as possible, and by a comparative study of the pat-
tern of the blood supply to the cerebrum, that is, the occurrence of the rete mirabile caroticum. He found morphological
indications of a carotid sinus, based on the presence of a dilation at the origin of the internal carotid artery and on innervation particularly by the glossopharyngeal nerve, in a great
many species. Furthermore, by a correlation of these findings
with those based on a study of embryological material for the

presence of a rete mirabile throughout the vertebrate series, h
showed that the rete occurred only in species in which blood wa
supplied to the brain at least in part by the external carotid
artery, and that, in such circumstances, the internal carotid
was not well developed. (Figure 6). Ask-Upmark believed that
the rete mirabile developes from the first or mandibular branchial arch and that it is homologous with the pseudobranch foun
in some fishes. In birds, reptiles and amphibians, no structure which could be considered to be genetically related to the
rete of mammals, was found. He believed that the rete mirabile
car̈ticum serves as a mechanical device to keep the pressure in
the arteries of the brain at a nearly constant level. He proposed the theory that it takes over the function of the carotid
sinus in the species in which the sinus is poorly developed.

The regulation of the blood supply to the brain has been
studied by two experimental methods. First, blood vessels have
been observed by means of a window in the cranium. Second,
blood pressure has been measured in the arteries and veins of
the brain. From the work of numerous investigators, Ask-Upmark
summarized the factors which determine the volume and rate of
blood passing through the brain. The first factor is the size
of the vascular bed, which may be varied by changes in the chemi
cal content of the blood, vasomotor impulses, and hydrodynamic
changes. The second, is the difference between the arterial an
venous pressures, which may be varied by changes in systemic
arterial blood pressure and in pressure of the cerebrospinal

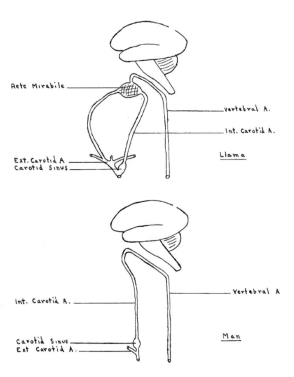

Rete Mirabile

Vertebral A.

Int. Carotid A.

Ext. Carotid A.
Carotid Sinus

Llama

Int. Carotid A.

Vertebral A

Carotid Sinus
Ext Carotid A.

Man

Rete Mirabile Caroticum

Fig. 6   After Ask-Upmark, 1935

fluid, as well as by obstructions of the venous outflow.

In view of these well established principles, Ask-Upmark undertook to determine the part played by the carotid sinus, in maintaining the cerebral blood supply. In anaesthetized cats, he stimulated the carotid sinus nerve and observed the caliber of the blood vessels in the brain, under a microscope attached to a window in the manner developed by Forbes. A constriction of the vessels occurred in most instances. As a rule, changes in caliber followed passively changes in the systemic blood pressure. Thus, when the sinus nerve was stimulated a reflex vasodilation occurred all over the body and was accompanied by a constriction of the pial arteries.

The hemodynamic effect might have obscured the presence of a reflex effect. Hence, to abolish the influence of changes in systemic pressure, Ask-Upmark first employed cross-circulation experiments and found no response of the cerebral vessels to stimulation of the carotid sinus nerve. Then he prepared neart-lung-head preparations in which the left ventricle was allowed to pump blood only to the head. There was no change in the diameter of the blood vessels in the majority of cases, although in four experiments stimulation of the sinus nerve did result in what was thought to be a reflex vasodilation.

These observations lend support to the present day concept that the state of tone of the cerebral vessels is reflexly maintained, but that the passive influence of changes in systemic blood pressure are more important in regulating the blood supply

27

Thus, when the carotid sinus mechanism elicits a reflex vasodilation, the fall in systemic blood pressure passively constricts the arterioles of the brain, thus preserving an adequate blood supply in the capillaries to furnish sufficient oxygen to the nervous tissues during the momentary duration of the reflex. They are also in agreement with the action of adrenalin which produces a general reflex vasoconstriction, in the cerebral vessels as well as elsewhere. However, the resulting rise in blood pressure is great enough to offset the reflex effect in the blood vessels of the cerebrum.

28

C.   Respiratory Reflexes

Almost from the time of the discovery of the aortic depressor
nerve, respiratory reflexes were known to accompany its stimulation. Electrical and mechanical stimuli were employed in the
earlier experiments. J. F. Heymans and C. Heymans (1927) were
the first to investigate the respiratory reflexes originating
at the level of tne aorta, by using a physiological stimulus.
They did this by means of the isolated head technique. The isolated head of dog B was perfused with blood from dog A. The
respiratory movements of the head of dog B were recorded. They
found that a fall of blood pressure in dog B, produced by with-
drawing a quantity of blood from the animal, brought about an
increase in the respiratory movements in the isolated head.
Furthermore, a rise in blood pressure in dog B, which was produced by either injecting blood or adrenalin into the animal,
brought about a decrease in respiratory movements in the isolated head. Thus the investigators concluded that a fall in the
systemic blood pressure produced hyperpnea and that a rise produced decreased respiration.

But the experiment described above did not localize the site
of the reflex. In order to do this, the investigators isolated
the head of dog B in the usual way and supplied it with blood
from dog A. Then they isolated the heart and lungs of dog B
and supplied them with blood from dog C. They found that a decrease in the blood pressure in the heart and lungs of dog B

produced an increase in respiratory movements in its isolated
head. An increase in blood pressure produced the opposite effect. Another experiment in which the lungs only of dog B,
with isolated head, were supplied with blood from dog C elimi-
nated the lungs as a source of the reflex and left the heart
and aorta as the only alternatives. That these structures contain the receptors of the respiratory reflex was shown in a
direct manner by a third experiment in which only the heart of
dog B, with isolated head, was supplied with blood by dog C.
The reflex was elicited by blood pressure changes in the usual
way.

As stated above in the introduction, physiologists realized
before the discovery of the carotid sinus reflex, that changes
in the pressure in the blood vessels of the head modified the
respiration. Except for Pagano (1900) and Siciliano (1900)
physiologists believed that these modifications were due to a
direct effect of variations in blood pressure on the medullary
centers. But after the discovery of the carotid sinus nerve,
Moissejeff (1926) demonstrated that the carotid sinus elicits
respiratory reflexes. Heymans (1929) confirmed this using the
blind sac technique employed by Moissejeff. Both investigators
found that the same response to an increased endosinusal pressure occurred as in the case of increased pressure in the heart
and aorta. Furthermore, these effects do not occur after sec-
tioning the carotid sinus nerves. Heymans and co-workers showed
that pressure changes within the physiological limits could e-
licit this reflex and that the reflex modification of respira-

tory movements were a true measure of pulmonary ventilation, by
an analysis of the respiratory volume. In order to determine
whether or not a part of the respiratory response was due to a
direct effect upon the centers, they exposed the vertebral arteries, cut the aortic depressor nerves and tied off the efferent arteries of the carotid sinus. Occlusion of the vertebral
arteries did not elicit a change in respiration. Thus the respiratory effects produced by altering the pressure of the
blood in the arteries of the head, were reflex in origin rather
than central.

Pressoreceptors might be expected to occur in the blood vessels of the lungs because the sixth branchial arch persists to
form the pulmonary arteries. Tello (1924) found that the left
depressor nerve innervated the sixth pulmonary arch in the rabbit embryo. However, Nonidez (1935) was not able to stain nerve
endings of the pressoreceptor type in the pulmonary arteries of
young cats, guinea pigs and rabbits.

As stated above, J. F. Heymans and C. Heymans obtained evidence that the pulmonary vessels did not contain receptors which
were capable of stimulation by changes in the blood pressure.
However, they found that Hering-Breuer reflexes occurred. The
impulses which initiate these reflexes arise from receptors
which are located in the alveolar ducts. Stretching of one set
of receptors due to inflation during inspiration inhibits the
respiratory center, and stretching of another group of receptors due to tne collapse of the lungs during forced expiration
stimulates the respiratory center. Apparently the receptors

giving rise to impulses which modify respiration occur in the
alveolar ducts and not in the blood vessels of the lungs.

In the embryo, the aorta and pulmonary artery are joined by
the duct of Botallo. "The ductus Botalli is important in the
embryonic circulation of amniotes, as the larger part of the
blood goes through it to reach the dorsal aorta, as long as the
allantois is the organ of respiration, while only enough blood
goes through the pulmonary artery to nourish the lung. With the
first inspiration of air, the duct closes and all the blood passing into tne last arch goes to the lung". (Kingsley, 1926)

In view of the importance of the duct of Botallo in the fetal
circulation one might expect to find vascular reflexes originating from it. That pressoreceptor nerve terminations occur here
has been demonstrated by Nonidez (1935) and Muratori (1937).
However, no physiological work demonstrating the presence of reflexes appears to have been done. Furthermore, Clark (1934) has
presented evidence that the pressoreceptors in the dog and cat
do not begin to function until a few days after birth, although
they are fully developed at term. Also, evidence to be dis-
cussed later in this paper in connection with the carotid body
indicates that the chemoreceptors in the fetus are non-function-
al, as well.

D. Phylogenetic Considerations

In the Mammalia, a carotid sinus has been demonstrated in
man, monkey, dog, cat, rabbit, rat, mouse, horse and ox. However, physiological experiments have not always been used to show

that the anatomical structure is a carotid sinus in the physiological sense. Ask-Upmark (1935) found anatomical evidence of
a carotid sinus in the opossum, flying fox, armadillo, camel,
llama and several other species which are less commonly subject-
ed to biological investigation. The aortic depressor nerve has
an equally widespread distribution among mammals.

In the Aves, not much work has been done. Ask-Upmark found
a segment of the carotid artery supplied by the glossopharyngeal
nerve in the swan, Stanley crane and Canadian goose. He believed that the specially innervated portion of the carotid was homologous with the carotid sinus in mammals. The observations of
other workers have been limited for the most part to the chick.

In the Reptilia, some evidence for a carotid sinus has been
obtained. Thus, Ask-Upmark found a segment of the carotid artery innervated by the glossopharyngeal and also the vagus, in the
alligator, turtle and snake. In the turtle, Kazem-Beck (1888)
stimulated the central end of a slender nerve which terminated
near the heart, in the aorta. He found that hypotension and
bradycardia resulted. Thus, there is evidence for an aortic
depressor mechanism in the turtle.

In the Amphibia, there is physiological evidence for both a
carotid sinus nerve and an aortic depressor nerve. Meyer (1927)
found that the glossopharyngeal nerve gave off small fibers to
the carotid gland in the frog. He stimulated the central end of
these fibers and obtained a fall in blood pressure. Nikiforowsky (1913) stimulated the central end of the cut vagus nerve in

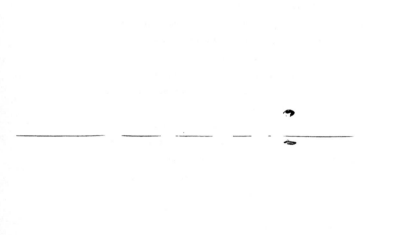

the frog. He observed a definite dilation of the blood vessels in the mesentery. In addition, he found a slowing of the heart and fall in blood pressure of 2 to 3 millimeters of water. Lutz and Wyman (1932b) found that reflex cardiac inhibition resulted from mechanical or electrical stimulation of the gills in the mud-puppy. Also, Lutz (1933) found that the response to adrenalin in the tropical toad suggested the existence of a reflex cardio-inhibitory mechanism stimulated by an increased intravascular pressure.

In the Pisces, there are found reflexes which control the heart rate and respiration. Lutz and Wyman (1932a) found that mechanical and electrical stimulation of widespread areas in the dogfish brought about cardiac inhibition. Slowing of the heart was obtained when mechanical stimulation was applied to the nasal openings, the sides of the fish, and region of the gills. Sensory endings were also found in the gill vessels since perfusion of the afferent branchial arteries with ureasaline or heparinized dogfish blood slowed the heart. "It is conceivable, therefore, that in the course of evolution the widespread sensory areas of the ancestral form, possibly typified by the elasmobranch, with Necturus as an intermediate type, were concentrated or restricted until the condition seen in the mammal was reached. Since the carotid arteries of the mammal are derivatives of the primitive branchial system, the reflex cardiac inhibition of branchio-vascular origin may exemplify the evolutionary forerunner of the carotid sinus reflex in mammals". (Lutz and Wyman, 1932)

34

Part II.   Reflex Responses to Changes
in
Chemical Composition of the Blood

Heymans and his collaborators (1933) demonstrated that the
aorta was the place of origin of reflexes initiated by changes
in the chemical content of the blood. By means of the isolated
head technique, they found that hyperventilation in dog B brought
about a decrease in respiratory movements of its isolated head.
On the other hand, asphyxia, produced by either stopping the arti-
ficial respiration or introducing carbon dioxide, brought about
an increase in respiratory movements. This reflex was not initiated by stretch receptors in the lungs since it was not abolished
by cutting the branch of the vagus innervating the lungs.

By such experiments it was shown that carbon dioxide lack
reflexly inhibits respiration, and that both oxygen lack and carbon dioxide excess reflexly augment respiration. The investigators made use of the same experiments which were used to localize
the pressure receptors, only in this instance they varied the
carbon dioxide and the oxygen content of the inspired air. The
chemical receptors were found to be limited to the region of the
heart and aorta. It had been known for some time that changes
in the carbon dioxide and oxygen content of the blood brought
about respiratory reflexes. These reflexes were ascribed to a
direct effect on the centers, as they were in the case of blood
pressure changes. But the previous establishment of the aorta
as a source of respiratory reflexes elicited by changes in the

chemical content of the blood suggested that the homologous carotid sinus might also be a source of these reflexes. Heymans and
his co-workers were the first to demonstrate that this was true.
As in the case of the aorta, perfusion with hypercapnic and anoxic blood reflexly augmented respiration. Furthermore, control
experiments in which the same perfusion fluid was allowed to
reach the medullary center showed that the effects were chiefly
reflex in nature. Schmidt (1932) confirmed Heymans with respect
to the existence of a reflex response to oxygen lack at the level
of the carotid sinus. However, he found that carbon dioxide excess had only a slight reflex effect and that it stimulated the
respiratory center directly.

Recent work by Comroe and Schmidt (1938) proved that the pressoreceptors and chemoreceptors concerned in the carotid sinus reflexes are automatically separate. The chemoreceptors were found
exclusively in the carotid body. Comroe (1938) found that the
aortic body also contained chemoreceptors.

A. The Carotid Body (Glomus caroticum)

The carotid body is a small globose structure which is located near the bifurcation of the common carotid artery. It has
been found in most of the higher vertebrates which have been examined for its presence. It is sometimes so inconspicuous and
small that it can not be seen macroscopically. However it is
quite noticeable in man; its size was reported by Luschka (1862)
as being 5 x 2.5 x 1.5 millimeters. Comroe and Schmidt (1938)

described the carotid body in the dog. In this animal, the carotid body can be seen by the naked eye as a reddish nodule on the medial side of a small artery which arises from either the external carotid artery or the occipital artery. This small artery which supplies the carotid body becomes veinlike in appearance but regains its arterial nature beyond the carotid body and anastomoses with other arterial branches in the neck. Apparently this artery is both the afferent and efferent supply, inasmuch as no venous channels have been found. The carotid body in the dog is innervated by the carotid sinus nerve.

Anatomists have long been interested in the microscopical constituents of the carotid body, chiefly in regard to the presence or absence of chromaffin tissue. According to Smith (1924) the carotid body is composed of three cellular elements. First, chromaffin cells are found in varying amounts in different species; thus, they are abundant in the cow, present more sparsely in the cat and absent in the rat. Their embryological origin is uncertain. Second, non-chromaffin cells similar to the chromaffin cells in size and shape, but not taking the chrome stain, are present. These cells are also of uncertain origin. Third, mesodermal constituents are found. They are developed from the mesenchyme of the third branchial arch.

Because of the presence of chromaffin tissue in the carotid body, it came to be known as a chromaffin body or paraganglion as proposed by Kohn (1900). However, this theory was not universally accepted. Smith (1924) believed that an analysis of

the growth changes in the region of the development of the carotid body would aid in explaining its origin and function. From a study of mammalian embryos (cat, rat, pig, sheep, cow) she presented four types of evidence for an origin from the third branchial arch. First, the carotid body developed near the bifurcation of the common carotid artery and the proximal part of the internal carotid artery developed from the third branchial arch of the embryo. Second, the carotid body is innervated by a characteristic branch of the glossopharyngeal nerve which is embryologically and comparatively the nerve of the third branchial arch. Third, the vascular supply has its anlage in the blood vessels of the third branchial arch. Fourth, the mesodermal constituents may be traced directly to the third arch region. The presence of vagus nerve fibers, autonomic nerve fibers and chromaffin tissue are explained by growth changes which bring them into the region of the carotid body at the time of its formation. Furthermore there has been no clean cut evidence for an endocrine function of this organ. Thus, the carotid body can not rightly be classified as a paraganglion. It is not homologous with the paraganglia of Zuckerkandl or the medulla of the suprarenal gland.

Boyd (1935) confirmed the opinion that the carotid body developed from the third branchial arch. He found that the carotid body was present in the human embryo at the 13 millimeter stage as a collection of cells around the third arch artery at the origin of the external carotid artery. At this early stage, the

glossopharyngeal nerve was associated with these cells. No sympathetic nerves were seen. At the 12 to 20 millimeter stage, the
colltction of cells lost its connection with the third arch
artery and increased in size possible due to the influx of nerve
fibers from the glossopharyngeal nerve, vagus nerve or superior
cervical ganglion.

The carotid body is highly vascularized and richly innervated.
De Castro (1926, 1928) from a study of the carotid body in man
and animals, reported that the cells were in contact with nerve
terminations on one side and capillaries on the other side. As
a result of his physiological experiments and because of the
characteristic nature of the nerve endings, he believed they
were sensory endings. Muratori (1932) found similar nerve terminations in the carotid body of the bird. Also, Nonidez (1935a)
stained nerve terminations of the sensory type in the carotid
body of the chick using a modification of the silver nitrate
method of Ramón y Cajal. He described the terminations as small
rings and reticulated clubs so delicate that their demonstration
required an "aprochromatic immersion objective of high numerical
aperature". The terminal rings and clubs were found among the
cells of the body and were probably chemoreceptors. (Figure 7).
In addition to the chemoreceptors he found nerve endings in the
walls of the blood vessels supplying the carotid body; these
terminations were of the pressoreceptor type. Other nerve fibers believed to belong to the autonomic nervous system were
stained; no function has yet been assigned to them.

Pressoreceptor Nerve Terminations
From externa of aorta of rabbit

Chemoreceptor Nerve Terminations
From aortic glomus of Kitten

Fig. 7   After Nonidez, 1935

39

The presence of chromaffin cells has led numerous investigators to search for an endocrine function of the carotid body.
But modern and delicate tests for adrenin have been negative.
Extirpation experiments and the injection of prepared extracts
have not yielded consistent results. Christie (1933) reviewed
the literature pertaining to a possible glandular role of the
carotid body and obtained additional evidence for such a function. He prepared an extract from a large tumor of the carotid
body in man. The extract had the following physiological pro-
perties. First, it produced rhythmic contractions and increased
the tone of the resting uterus. Second, it produced a transient
fall in blood pressure when injected intravenously into a cat.
Third, it produced an acceleration of the heart. Christie also
found that atropine did not destroy the vasodepressor action of
the extract. Therefore he reasoned that the active ingredient.
was not acetyl choline. Furthermore, he found that an injection
of tissue extracts containing histamine did not produce results
similar to those of the tumor extract; consequently he believed
that the active principle in the extract was not histamine. He
proposed the name "carotidin" for this vasodepressor substance.

Christie (1933) also prepared extracts of a structure found
in the posterior cardinal sinus in elasmobranchs. He referred
to this structure as the carotid gland. He found evidence for
two kinds of secretion; first, a vasodepressor substance similar to that obtained from the tumor of the carotid body, and
second, a vasopressor substance similar to adrenalin. Christie

has stated that the inclusion of chromaffin tissue, which is scattered all along the sympathetic chain in elasmobranchs, in the carotid gland accounts for the vasopressor action. However, it is doubtful that this structure is homologous with the carotid body of mammals. Probably it is the anterior chromaphil body whose functions were investigated by Lutz and Wyman (1927) although Christie does not mention their work.

But even if the possibility of a secretion during pathological conditions be admitted, it is still more doubtful that one is elicited reflexly. Thus, although Ara (1934) was able to produce blood pressure changes in the dog upon injection of an extract prepared from fifty carotid bodies of cows, he could not demonstrate a secretion during elicitation of the carotid sinus reflex in the dog.

It was suggested by Dreuner (1924) that the carotid body might contain the receptors for the carotid sinus reflex. But numerous investigators have since shown, both anatomically and physiologically, that the pressoreceptors are found in the carotid sinus (Hering, 1927; Koch, 1931; Schmidt, 1932; Nonidez, 1935 and others). De Castro (1926) was the first to suggest that the carotid body might function as a receptor for impulses produced by changes in the chemical content of the blood. Bouckaert, Dautrebande and Heymans (1931) obtained evidence that the chemical receptors involved in the carotid sinus reflex are located in the carotid body. Thus, when the carotid sinus nerve was cut at the level of the bifurcation of the common carotid

artery, no response occurred to an increased intrasinusal pressure, but a response occurred to an injection of lobeline. On the other hand when the carotid sinus nerve was cut 0.5 to 1 centimeter above the bifurcation, no response occurred to either chemical stimuli or hydraulic stimuli. Although this experiment was inconclusive, since the medullary centers were not prevented from receiving the blood with altered chemical content, the opinion advanced by Heymans and co-workers has been supported by the experiments of other investigators (Schmidt, 1932; Samaan and Stella, 1935; Comroe and Schmidt, 1938; Bernthal, 1938).

Comroe and Schmidt (1938) upon completely isolating the carotid body receptors and the carotid sinus receptors in the dog found that lobeline produced hyperpnea when injected into the carotid body but had no effect upon the carotid sinus. Thus the reflexes initiated in the region of the carotid sinus by the injection of pharmacological agents (lobeline, nicotine, cyanide and so forth) and by changes in carbon dioxide and oxygen content of the blood originate in the carotid body and not the carotid sinus. The same investigators also found that changes in the perfusion pressure elicit no reflexes in the carotid body, implying that the pressoreceptors are limited to the carotid sinus. However, this latter point needs confirmation since de Castro (1928) obtained evidence that pressoreceptors in the carotid body are activated by an increased perfusion pressure when the carotid sinus is denervated. Both de Castro and Nonidez (1935a) have stained endings of the pressoreceptor type in

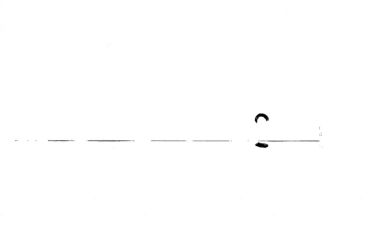

the blood vessels within the carotid body.

Comroe and Schmidt (1938) have also investigated the direct
effect of carbon dioxide tension and oxygen tension on the respiratory center in the medulla. Furthermore, they compared the
direct effect of carbon dioxide excess and oxygen lack with the
reflex effect in the same dog. This was accomplished by allow-
ing an anaesthetized dog to breathe a mixture of 10 per cent
carbon dioxide in oxygen and then pure nitrogen. The respiratory rate was recorded by means of a pneumograph and the volume
of expired air was measured. During this part of the experiment the carotid bodies were isolated from the vascular system
by perfusion with Locke's solution at a pressure above the systemic pressure of the blood. While the central effects were
occurring, hypercapnic and then anoxic blood was withdrawn from
the femoral artery and later used to perfuse the isolated carotid body of the same dog. Thus records of respiratory movements, respiratory minute-volume and blood pressure could be
compared with respect to the effects of anoxic and hypercapnic
blood within the physiological limits, upon the peripheral receptors and the medullary centers of the same experimental animal. It was found that the chemoreceptors respond to both oxygen lack and carbon dioxide excess but that the hyperpnea of increased carbon dioxide tension is much more central than reflex
and the hyperpnea of decreased oxygen tension is mainly or en-
tirely reflex.

Smyth (1937) investigated the central role of carbon dioxide

in regulating respiration, by means of chronic experiment performed on the rabbit. He defined a chronic experiment as "one
in which the respiratory reflexes are studied only after the
animal has completely recovered from the shock of the operation
for denervating the vaso-sensory areas". He found that a rabbit
with carotid sinus nerves and aortic depressor nerves cut responded to carbon dioxide excess in the same way that a normal
animal responded. This is further evidence that carbon dioxide
acts centrally and that peripheral receptors are of secondary
importance in respiratory changes brought about by increased
carbon dioxide tension in the blood.

Heymans (1933) believed that reflex vasomotor reactions to
variations in the chemical content of the blood were elicited
in the carotid body. However, Comroe and Schmidt (1938) found
that the vasomotor responses to alterations in the chemical
content of the blood were variable and probably secondary to the
respiratory responses. Bernthal (1938) perfused the isolated
carotid body of a dog with blood containing different chemicals
and varying carbon dioxide and oxygen tensions. Volume changes
in the axillary artery were recorded by means of the thermo-
electric method. It was found that anoxemia produced abrupt
and intense vasoconstriction and that hypercapnia sometimes
produced vasodilation. Thus there is evidence for reflex regulation of blood pressure as well as of respiration by the
chemical state of the blood perfusing the carotid body.

Bogue and Stella (1935) recorded action potentials from the

carotid body of the cat after removal of the pressure receptors by stripping the common carotid artery at its bifurcation. These potentials were not caused by changing the endosinusal pressure. They were different in nature from those previously obtained from the stretch receptors in the carotid sinus. The action potentials occurred during anoxemia and during injection of nicotine and cyanide. Samaan and Stella (1935) found that similar potentials occurred under normal conditions of respiration. They believed that the chemical receptors in the carotid body responded to either the carbon dioxide tension or the hydrogen ion concentration of the blood. As a result of their work they stated that some of the receptor endings had a low threshold and were continually discharging, thus contributing to the tonic activity of the respiratory center. The response occurring to oxygen lack was independent of the carbon dioxide content. However, the type of discharge in anoxemia and hypercapnia was the same and thus it was not known whether there were two distinct sets of chemical receptors, one for carbon dioxide and one for oxygen, or whether there was a single set.

The most recent work on the carotid body refutes the role of the hydrogen ion concentration of the blood as a stimulus for the chemical receptors. Thus Comroe and Schmidt (1938) found that changes in the hydrogen ion concentration of the blood during perfusion of the isolated carotid body did not alter the respiratory minute-volume. They favor the opinion advanced by Winder (1937) that the adequate stimulus is some disturbance in

the oxidative mechanism. Winder compared the reflex responses to carbon dioxide excess and oxygen lack in the carotid sinus region before and after local poisoning with monoiodoacetic acid which suppresses anaerobic glycolysis in muscle. During the poisoning there was a progressive increase in ventilation. After the poison had been effective, anoxic blood would not produce reflex hyperpnea but hypercapnic blood would do so. Hence it was concluded that anoxic stimulation of chemoreceptors was initiated by the intermediary or final products of glycolysis and that hypercapnic stimulation occurred by a different mechanism. This experimental work supports the theory of Gesell (1925) that the chemical control of respiration and circulation is effected by an alteration of the intracellular C_H of the nerve cells. This theory was originally applied to the neurones in the medullary centers. It has now been extended by Winder to the peripheral receptors in the carotid body.

It might be expected that the chemoreceptor reflexes would be developed in the fetus in view of the evidence against the existence of early proprioceptive reflexes from the carotid sinus and the obvious need for an adequate oxygen supply. Snyder and Rosenfeld (1937) observed the intrauterine respiratory movements of the fetus (rabbit, cat and guinea pig). They inhibited labor in the pregnant animal by an intravenous injection of pregnancy urine extract which did not interfere with fetal respiration. The following results were obtained. First, a lack of oxygen in the mother depressed or abolished the respiratory

movements in the fetus. Second, an excess of carbon dioxide in
the mother had little or no effect on the respiratory movements
in the fetus. Third, a lack of carbon dioxide in the mother
resulted in a depression of respiration or apnea in the fetus.
The effects on the respiration of the mother were the usual ones
obtained during a similar alteration of the chemical content of
the blood. Wright (1936) reported that in the denervated adult
mammal there was a respiratory depression following anoxemia.
Thus the fetus responded to anoxemia in a way which is much like
an adult whose carotid body has been denervated.

Among the Mammalia, a carotid body has been demonstrated in
numerous species (man, monkey, dog, cat, rabbit, mouse, pig,
sheep and cow). Ask-Upmark (1935) demonstrated that it is present in some of the less common mammals such as certain species
of Marsupialia and Insectivora. Among the Aves, a carotid body has been demonstrated although observations seem to have been
limited to a few species. Nonidez (1935_a) found a carotid body
with characteristic nerve endings in the chick. Ara (1934)
stimulated the carotid body in the hen and found that carotid
sinus reflexes resulted. In the Reptilia, not much is known of
the existence of the carotid body. Ask-Upmark (1935) has reported a small yellowish gland in the bifurcation of the common
carotid artery in a species of the Ophidiae. In the Amphibia,
it is problematical that a structure homologous with the carotid
body of mammals exists. Smith (1924) summarized the evidence
for and against considering the epitheliod bodies in amphibians

as homologous with the carotid body in mammals, but this was long before the function of the carotid body was known. In the Pisces, it is also problematical that a structure homologous with the carotid body of mammals exists, unless the structure described by Christie (1935) is a true homologue. However, Lutz and Wyman (1932) found that respiratory reflexes accompanied the cardio-inhibitory reflex induced by alterations in the pressure of the fluid which was used to perfuse the gill arches of the dogfish. No experiments using anoxic and hypercapnic blood as a perfusion fluid have as yet been performed in the elasmobranch.

The role of oxygen tension in eliciting reflexes of the carotid body can be understood best when the carotid sinus mechanism is examined from an evolutionary point of view. It is generally conceded that the respiratory center in mammals is autonomus. Lutz (1930) has shown that this is also true in the elasmobranchs. The peripheral receptors in mammals function chiefly under emergencies such as oxygen lack, abnormal changes in the chemical content of the blood and variation of blood pressure. In the elasmobranch, the need for these protective reflexes is greater because of the more direct exposure of the animal to the external environment and because of a less perfectly adjusted homeostasis. Perhaps this need accounts for the low threshold and wide-spread distribution of the receptive area for respiratory and cardiac reflexes in the elasmobranch. Evidently the chemoreceptors of the carotid body are the survi-

vals of a more wide-spread condition in the embryo or primitive
dogfish-like ancestor in the same way that the pressoreceptors
are survivals.

B.   The Aortic Body (Glomus Aorticum)

Other masses of tissue having a similar appearance to that
of the carotid body have been found near the aorta and heart.
Busacchi (1912) described in man such a structure which he
named the paraganglion cardioaortico superiore. Muratori (1934)
also found the same structure in man and emphasized its similarity to the carotid body. Nonidez (1935) discovered an homologous organ in the cat, rabbit and guinea pig; he named it the
aortic body because of its similarity to the carotid body. The
aortic body is paired. The left aortic body is situated among
the branches of the left aortic depressor nerve near its distribution to the aorta. The right one is situated among the
branches of the right aortic depressor nerve near its distribution to the right subclavian artery. The left aortic body receives its blood supply from a small branch of the aorta. The
right one receives its supply from a minute branch of the right
subclavian artery.

As in the case of the carotid glomus, the amount of chromaffin tissue present varies with the species; and its occurrence
is of secondary importance. The epitheliod type of cell con-
stituting the aortic body is of more significance than the
chromaffin cells. Nonidez (1937) found that many of the nerve

fibers in the aortic body terminated in minute rings and club shaped endings similar to those of the carotid body. He believed that these nerve endings were chemoreceptors. In addition he described nerve endings of the pressoreceptor type in the walls of the arterioles within the aortic body. (Figure 7, page 38).

The position and structure of the aortic body suggests that it is developed from the fourth branchial arch. And recent physiological experiments demonstrate that it is concerned in the aortic depressor reflexes. Thus, Comroe (1938) localized the aortic arch chemoreceptors in the aortic body. He found that vasomotor and respiratory reflexes occurred when lobeline and sodium cyanide were injected into the aorta near the point of origin of the small artery which supplied the aortic body. Furthermore, stimulation of the branch of the vagus nerve innervating the aortic body produced hypertension and hyperpnea. But the vasomotor responses were more marked than the respiratory responses. Hence, Comroe believed that the aortic body was concerned primarily with blood pressure and that the carotid body was concerned chiefly with respiration.

The aortic body has been found in certain species of Mammalia (man, cat, rabbit and guinea pig). In the Aves there is some evidence for the existence of an aortic body. Palme (1934) described a structure in the finch, which he called the paraganglion supracardiale superius. Nonidez (1935a) believed that the paraganglion found by Palme, was homologous with the aortic bo-

dy of mammals. Nonidez discovered encapsulated corpuscles in
the wall of the aorta in the chick. These corpuscles contained
cells and nerve endings of the type found in the mammalian
aortic body.

C.   The Pulmonary Body (Glomus Pulmonare)

In addition to describing the aortic body in man, Busacchi
(1912) mentioned a similar structure in the space between the
aorta and pulmonary arteries just above the semilunar valves.
Muratori (1934) also found this structure in man and called it
the paraganglion carioaortico inferiore. Nonidez (1936) found
an homologous structure in the cat and he named it the paraganglion aorticum. It received its blood supply from a small
branch of the pulmonary artery. It was innervated by nerve
fibers branching from the right cardiac nerve and right vagus,
near the point of origin of the recurrent laryngeal.

The position and structure of the pulmonary body suggests
that it is developed from the sixth branchial arch. However,
no physiological function has yet been attributed to it. J. F.
Heymans and C. Heymans (1925) perfused the isolated lungs of a
dog B, with blood from another dog C. When tney produced an
asphyxiation in dog C and consequently an oxygen lack in the
blood vessels in tne lungs of dog B, no modifications of the
respiration of dog B occurred. Thus, the investigators concluded that the pulmonary vessels did not contain receptors which
were capable of stimulation by the chemical composition of the

blood.

The pulmonary body has been found in a few species of Mammalia (man, and cat). In the Aves, there is evidence for the existence of a pulmonary body. Thus, Palme (1934) found a structure in the finch which he named the paraganglion supracardiale inferius. Nonidez (1936) believed that this structure, which was found by Palme, was homologous with the pulmonary body of mammals.

The recent findings concerning the function of the carotid and the aortic bodies lead to speculation concerning the possible function of the coccygeal gland, which was described in man by Luschka. Its anatomical structure resembles that of the carotid body. Application of the silver nitrate stain method and the perfusion experiments to the coccygeal gland might contribute to the knowledge of its function.

SUMMARY

The evidence that the carotid sinus and its homologues are part of an evolutionary mechanism in summarized below.

1. Comparative anatomical investigations have demonstrated the following points.

(a) The carotid sinus and its homologues are found in many of the higher vertebrates.

(b) They are innervated by gill arch nerves.

(c) The nerve terminations are characteristic throughout the vertebrate series.

2. Comparative embryological investigations have demonstrated that the carotid sinus and its homologues are developed from the branchial arch vessels of the embryo.

3. Comparative physiological investigations have demonstrated that the carotid sinus and its homologues contain pressoreceptors and chemoreceptors.

(a) The pressoreceptors are important in regulating the heart and respiration in aquatic vertebrates. They also take part in regulating the heart, blood pressure and, to a less extent, respiration in the air-breathing vertebrates.

(b) The chemoreceptors are important in protecting the aquatic vertebrates from anoxemia by regulating the heart and respiration. They also take part in regulating the heart, blood pressure and respiration in the air-breathing vertebrates.

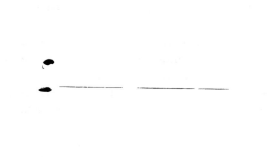

BIBLIOGRAPHY

Anrep, G. V. and H. N. Segall, 1926. The central and reflex
  regulation of the heart rate. J. Physiol., 61, 215.

Ara, G., 1934. Ricerche sulla funzione del glomo carotideo.
  Arch. Fisiol., 33,325. Quoted by Young, J Z. Physiol.
  Abstr., 19,483.

Ara, G., 1934. Il riflesso di Pagano-Hering nel pollo. Arch.
  Fisiol. 33,632. Quoted by Young, J. Z. Physiol. Abstr.,
  19,452.

Ara, G. and U. Sacchi, 1932. La presenza di una sostanza
  ipertensiva nel glomo carotideo. Boll. Soc. ital. Biol.
  spec., 7, 351. Quoted by Young, J. Z., Physiol. Abstr.,
  18,107.

Ara, G. and U. Sacchi, 1934. La presenza di una sostanza
  ipertensiva nel glomo carotideo. Arch. Fisiol., 33,307.
  Quoted by Young, J. Z. Physiol. Abstr., 19,483.

Ask-Upmark, E., 1935. The Carotid Sinus and the Cerebral Circulation. Levin and Munksgaard, Copenhagen.

Bainbridge, F. A., 1915. The effect of venous filling upon
  the rate of the heart. J. Physiol., 50,65.

Berntnal, T., 1938. Chemo-reflex control of vascular reactions through the carotid body. Am. J. Physiol. 121, 1.

Bogue, J. Y. and G. Stella, 1935. Afferent impulses in the
  carotid sinus nerve (nerve of Hering during asphyxia and
  anoxemia. J. Physiol., 83,459.

Bouckaert, J. J., L. Dauterebande and C. Heymans, 1931. Dis-
  sociation anatomo-physiologigue des deux sensibilités du

sinus carotidien: sensibilité à la pression et sensibilité chimique. Ann. Physiol. Physioch., 7,207.

Boyd, J. D., 1965. The development of the human carotid body.
Proc. Am. Assoc. Anat., Anat. Rec., 61, Suppl., 52.

Bronk, D. W., L. K. Ferguson and D. Y. Solandt, 1934. Inhibition of cardiac accelerator impulses by the carotid sinus.
Proc. Soc. Exp. Biol., 31,579.

Bronk, D. W. and G. Stella, 1932. Afferent impulses in the
carotid sinus nerve. I. The relation of the discharge
from single end organs to arterial blood pressure. J.
Cell. and Comp. Physiol., 1,113.

Bronk, D. W. and G. Stella, 1932. Afferent impulses from single
end organs in the carotid sinus. Proc. Soc. Exp. Biol.
and Med., 29,443.

Bronk, D. W. and G. Stella, 1935. The response to steady pressures of single end organs in the isolated carotid sinus.
Am. J. Physiol. 110,708.

Busacchi, P., 1912. I corpi cromaffini del cuore umano. Arch.
ital. di anat. e di embriol., 11,352. Quoted by Muratori,
1937.

Castro, F. de, 1926. Sur la structure et l'innervation de la
glande intercarotidienne (glomus caroticum) de l'homme et
des mammifères et sur un nouveau système d'innervation
autonome du nerf glossopharyngien. Trav. Labor. Rech.
Biol. Univ. Madrid, 24,365.

Castro, F. de, 1928. Sur la structure et l'innervation du si-

nus carotidien de l'homme et des mammifères. Nouveaux
faits sur l'innervation et la fonction de glomus caroticum. Trav. Labor. Rech. Biol. Univ. Madrid, 25,331.

Christie, R, V., 1933. The function of the carotid gland (glomus caroticum) I. The action of extracts of a carotid
gland tumor in man. Endocrinology, 17,421.

Christie, R. V., 1933. The function of the carotid gland. II.
The action of extracts of the carotid gland of the elasmobranch. Endocrinology, 17,433.

Clark, G. A., 1934. The development of blood pressure reflexes.
J. Physiol., 83,229.

Code, C. F., W. T. Dingle and V. H. K. Moorhouse, 1936. The
cardiovascular carotid sinus reflex. Am. J. Physiol.,
115,249.

Comroe, J. H., 1938. Localization and physiological significance of the aortic chemoreceptors in the dog. Proc.
Am. Physiol. Soc., 41.

Comroe, J. H. and C. F. Schmidt, 1938. The part played by reflexes from the carotid body in the chemical regulation
of respiration in the dog. Am. J. Physiol., 121,75.

Cyon. E. de and Ludwig, 1866. Ber. über die Verh. d. Konigl.
Sachs. Ges. d. Wiss., Leipzig (Math.-phys. Kl) 18,307.
Quoted by Heymans, C., J. J. Bouckaert and P. Regniers,
1933.

Druener, 1924. Über die anatomischen Unterlagen der Sinus-
reflex Herings. Dtsch. med. Wochscnr., 51,559. Quoted

by Ask-Upmark, 1935.

Evans, C. Lovatt, 1936. Recent Advances in Physiology. Blackston, Philadelphia.

François-Franck, 1877. Trav. lab. Marey, 3,273. Quoted by Heymans, C., J. J. Bouckaert and P. Regniers, 1933.

Gesell, R., 1925. The chemical regulation of respiration. Physiol. Rev. 5,551.

Gesell, R. and C. Moyer, 1937. Factors which determine the rate and depth of breathing. Am. J. Physiol., 119,55.

Haller, 1766. Elementa Physiologiae Corporis Humani. Lausannae.

Hédon, 1910. Arch. intern. physiol. 10,192. Quoted by Heymans, C., J. J. Bouckaert and P. Regniers, 1933.

Hering, H. E., 1924. Pflügers Arch., 206,72. Quoted by Heymans, C., J. J. Bouckaert and P. Regniers, 1933.

Heymans, C., J. J. Bouckaert and P. Regniers, 1933. Le Sinus Carotidien et la Zone Homologue Cardio-aortique. G. Dion, Paris.

Heymans, C. and A. Ladon, 1924. Perfusion et survie de la tête sectionée du chien. Anémie bulbaire, automatisme respiratoire. C. R. Soc. Biol., 90,93. Quoted by Heymans, C., J. J. Bouckaert and P. Regniers, 1933.

Heymans, J. F. and C. Heymans, 1925. Sur le mécanisme de l'apnée réflexe ou pneumogastrique. C. R. Biol., 92,1335. Quoted by Heymans, C., J. J. Bouckaert and P. Regniers, 1933.

57

Kazem-Beck, 1888. Arch. f. Anat., 325.

Kingsley, J. S., 1926. Outlines of Comparative Anatomy of
Vertebrates. Blakiston, Philadelphia.

Koch, E., 1931. Die reflektorische Selbststeuerung des Kreis-
laufes. Theodor Steinkopff, Dresden and Leipzig.

Konn, A., 1900. Über den Bau und die Entwicklung der sogenan-
nten Carotisdrüse. Arch. f. mikr. Anat., 56,81. Quoted
by Smith, 1924.

Luschka, H., 1862. Über d. drüsenartige Natur d. sog. Ganglion
intercaroticum. Archiv. f. Anat., Physiol. u. wissensch.
Med. Herausgeg. v. Reichert und DuBois-Reymond, Leipzig.
Quoted by Ask-Upmark, 1935.

Lutz, B. R., 1930. Respiratory rhythm in the elasmobranch
Scyllium canicula. Biol. Bull., 59,179.

Lutz, B. R., 1933. The effect of adrenalin chloride and toad,
venom on the blood pressure and heart rate of the tropical toad, Bufo marinus. Biol. Bull., 64,299.

Lutz, B. R. and L. C. Wyman, 1927. The chromaphil tissue and
interrenal bodies of elasmobranchs and the occurrence of
adrenin. Jour. Exp. Zool., 47,295.

Lutz, B. R. and L. C. Wyman, 1932a. Reflex cardiac inhibition
of branchiovascular origin in the elasmobranch Squalus
acanthias. Biol. Bull. 62,10.

Lutz, B. R. and L. C. Wyman, 1932b. The evolution of a carotid
sinus reflex and the origin of vagal tone. Science, 75,
590.

McDowall, R. J. S., 1935. A cardio-pressor nerve. J. Physiol.
    83, Proc. Physiol. Soc., 37.

McDowall, R. J. S., 1935. The nervous control of the blood
    vessels. Physiol. Rev., 15,98.

Meyer, F., 1927. Versucke über Blutdruckzügler beim Frosch.
    Pflügers Arch., 215,545.

Moissejeff, E., 1927. Zur Kenntnis des Carotissinusreflexes.
    Zeitschr. f. exper. Med., 53,696. Quoted by Heymans, C.,
    J. J. Bouckaert and P. Regniers, 1933.

Muratori, G., 1932. Ricerche istologiche e sperimentali del
    tessuto paragangliare, annesso al sistema del vago. Boll.
    Soc. ital. Biol. sper., 7,137.

Muratori, G., 1937. Aspetti istologici dell innervazione e
    significato embriologico della sede delle zone arteriose
    pressorecettrici negli amnioti e nell'uomo. Monitore
    Zoologico Italiano, 67,228.

Nash, R. A., 1926. Concerning the part played by the sinus
    caroticus in the central regulation of the circulation.
    J. Physiol. 61, Proc. Physiol. Soc., 28.

Nikiforowsky, P. M., 1913. J. Physiol. 45,459.

Nonidez, J. F., 1935. The aortic (depressor) nerve and its
    associated epithelioid body, the glomus aorticum.
    Am. J. Anat., 57,259.

Nonidez, J. F., 1935a. The presence of depressor nerves in
    the aorta and carotid of birds. Anat. Rec., 62,47.

Nonidez, J. F., 1936. Observations on the blood supply and the

innervation of the aortic paraganglion of the cat.
J. Anat., 70,215.

Nonidez, J. F., 1937. Identification of the receptor areas in
the venae cavae and pulmonary veins which initiate reflex
cardiac acceleration (Bainbridge's reflex). Am. J. Anat.,
61,203.

Pagano, G., 1900. Arch. ital. de Biol., 33,1. Quoted by
Ask-Upmark, 1935.

Palme, F., 1934. Die Paraganglion über dem Herzen und im End-
igungsgebiet des Nervus depressor. Zeit. f. mikr. anat.
Forsch, 36,391. Quoted by Nonidez, 1935.

Samaan, A. and G. Stella, 1935. The response of chemical receptors of the carotid sinus to the tension of CO_2 in the
arterial blood in the cat. J. Physiol. 85,309.

Schäfer, 1877. Über die aneurysmatische Erweiterung der Ca-
rotis interna an inrem Ursprung. Zschr. f. Psych.,
34,438. Quoted by Ask-Upmark, 1935.

Schmidt, C. F., 1932. Carotid sinus reflexes to the respiratory center. Am. J. Physiol. 102,94.

Schmidt, C. F., 1938. Part IV. The Respiration. MacLeod's
Physiology in Modern Medicine. C. V. Mosby, St. Louis.

Siciliano, 1900. Arch. ital. de Biol., 33,338. Quoted by
Ask-Upmark, 1935.

Smith, C., 1924. The origin and development of the carotid
body. Am. J. Anat. 34,87.

Smyth, D. H., 1937. The study of the carotid sinus respiratory

reflexes by means of chronic experiments. J. Physiol.
88,425.

Snyder, F. F. and M. Rosenfeld, 1937. Direct observation of
intrauterine respiratory movements of the fetus and the
role of carbon dioxide and oxygen in their regulation.
Am. J. Physiol., 119,153.

Tello, J. F., 1924. Trav. labor. rech. biol., Univ. Madrid.
22,295. Quoted by Heymans, C., J. J. Bouckaert and P.
Regniers, 1933.

Thomas, C. B. and C. M. Brooks, 1937. The effect of sympathec-
tomy on the vasomotor carotid sinus reflexes of the cat.
Am. J. Physiol., 120,195.

Tschermak, J., 1866. Über mechanische Vagusreizung beim Men-
schen. Jena Z. Med. u. Naturwiss, 2,384. Quoted by Heymans, C., J. J. Bouckaert and P. Regniers, 1933.

Winder, C. V., 1937. Pressoreceptor reflexes from the carotid
sinus. Am. J. Physiol. 118,379.

Winder, C. V., 1937. On the mechanism of stimulation of carotid
gland chemoreceptors. Am. J. Physiol., 118,389.

Wright, S., 1936. Further observations on mode of action of
oxygen lack on respiration. Quart. J. Exper. Physiol.,
26,63.